Daiane Corrêa

FOLIAR FERTILISERS AND BIOSTIMULANTS

Daiane Corrêa

FOLIAR FERTILISERS AND BIOSTIMULANTS

Effects on agriculture

ScienciaScripts

Imprint
Any brand names and product names mentioned in this book are subject to trademark, brand or patent protection and are trademarks or registered trademarks of their respective holders. The use of brand names, product names, common names, trade names, product descriptions etc. even without a particular marking in this work is in no way to be construed to mean that such names may be regarded as unrestricted in respect of trademark and brand protection legislation and could thus be used by anyone.

Cover image: www.ingimage.com

This book is a translation from the original published under ISBN 978-620-2-80852-1.

Publisher:
Sciencia Scripts
is a trademark of
International Book Market Service Ltd., member of OmniScriptum Publishing Group
17 Meldrum Street, Beau Bassin 71504, Mauritius
Printed at: see last page
ISBN: 978-620-3-36343-2

FOLIAR FERTILIZERS AND BIOSTIMULANTS
Effects on agriculture

DAIANE CORRÊA

Agronomist
Dr. Plant Production

SUMMARY

Chapter 1

FOLIAR FERTILIZER IN PLANT DEVELOPMENT

Daiane Corrêa

Foliar fertilizers and plant biostimulants

Foliar fertilizers and biostimulants are defined as synthetic compounds or bioregulators, added with other substances, such as nutrients, growth hormones, amino acids, humic substances, organic substances, macro and micro nutrients, mineral salts, microorganisms, carbohydrates and algae extracts (DABADIA, 2015; PETRI et al., 2016).

The use of foliar fertilizer and biostimulants as an agronomic technique aims to promote, modify or inhibit the metabolic and physiological processes of plants, through the stimulation of cell division, cell differentiation and effective fruiting, to optimize agricultural production in some crops, which has been gaining space in recent years, being used to enhance the productive performance of agricultural crops (DOURADO NETO et al., 2004).

Foliar fertilizers and biostimulants act on the physiology of the plant in different ways and by different routes to improve productivity and quality. They are products of various origins, without residues, and are increasingly used in agriculture. Many of the beneficial effects of biostimulants, are correlated with their ability to influence the hormonal activity of plants, which is responsible for regulating their development, as well as producing responses to the environment where they are found (DANTAS et al., 2012).

Among the known plant growth regulators are auxins, gibberellins, cytokinins, abscisic acid, ethylene, brassinosteroids, jasmonates, and salicylates. Among the main hormones that constitute commercial biostimulants are auxins, gibberellins and cytokinins, which can be added to foliar fertilizers, as well as humic acids, fulvic acids and organic carbon (ALBUQUERQUE et al., 2008).

Auxins are plant regulators that act on the rooting of plants, promoting greater growth of the root system. Auxin participates in the regulation of apical dominance, lateral root initiation, leaf abscission, vascular differentiation, floral bud formation, and fruit development (TAIZ; ZEIGER, 2013), also forming new meristematic centers or activating existing meristems that induce root formation (PETRI et al., 2016).

Gibberellins are present throughout the plant and act during the entire vegetative and reproductive cycle. They are functional in plant physiological processes related to stem elongation, mobilization of reserves, flower induction, fruit setting and growth, and induction of seed germination (ALBUQUERQUE et al., 2008).

Gibberellic acid is associated with the promotion of stem growth and the application of this plant regulator to the intact plant can induce a significant increase in its height. The exogenous application of gibberellin promotes the elongation of the internodes and associated with this effect, there is also a reduction in stem thickness and leaf size, in addition to the light green coloration of the leaves (TAIZ; ZEIGER, 2013).

Cytokinins have shown effects on developmental physiological processes, participating in the regulation of many plant tissue processes, including leaf senescence, nutrient mobilization, apical dominance, apical meristem formation and activity (ALMEIDA et al., 2016), as well as floral development, seed germination, and bud dormancy breaking. It can also mediate actions involved with plant development regulated by luminosity, including the differentiation of chloroplasts, the development of autotrophic metabolism and the expansion of leaves and cotyledons (BOURSCHEIDT, 2011).

Leaves, like roots, also exhibit the ability to absorb different biostimulant plant hormones composed of nutrients, through the penetration of the substances into the lipid cuticle, with uptake through the plasma surface and passage through the plasma membrane to enter the cytoplasm of the cells present in the leaf (TAIZ; ZIEGER, 2013).

LIMA et al. (2011), found that the amounts of macronutrients that should be applied via soil, are much greater than those applied via foliar, demonstrating greater efficiency of use, however, foliar fertilization should be conditioned as a complementary practice of fertilization performed in the soil, in which macro and micronutrients are supplementary components to the development of plants.

The foliar fertilizers available on the market have the objective of supplying one or more elements essential to the development of the plant, applied to the aerial part of the plant. The application of foliar fertilizers, with K, Ca, Mg, Mn, B and Z in their composition, has been adopted by some agricultural producers, aiming to increase production and quality, and foliar fertilizers composed of salts and chelates are also used (MELLO; VITTI, 2002; GENUNCIO et al., 2010).

During the initial phase of development of the germination process and plant emergence, the treatment with phytosanitary products is widespread, in which insecticides and fungicides are used in seed treatment. Besides these, other products that alter the growth and development of plants can be used, such as growth regulators, complexes with macro and micronutrients, together or separately with biostimulants (BINSFELD et al., 2014).

The adoption of foliar fertilizers through seed treatment, act of promoting root development of seedlings, favoring better characteristics in seedling production and consequently, improving the vegetative development of different agricultural species (HORNKE et al., 2018).

The application of foliar fertilizers and biostimulants composed of foliar fertilizers in the initial stages of plant development may also confer greater resistance to the attack of insect pests, diseases and nematodes in the midst of the crop. Thus, the faster and more uniform establishment of the plants results in good performance in the absorption of nutrients and, consequently, in their productive potential (FAIAD et al., 2002).

In this context, foliar fertilizers and plant biostimulants have the ability to provide greater absorption of water and nutrients, providing greater growth and structuring of the plants and consequently greater resistance to adverse conditions during critical periods during the crop cycle (RODRIGUES et al., 2015).

Nutrition of vegetable plants

Vegetables are very demanding plants in macro and micronutrients, and the contents and accumulation of elements by the crop may vary, mainly according to the stage of development of the plant, the cultivar and the estimated production that is desired (BASTOS et al., 2013).

 Soil fertility and crop requirements are extremely important factors in cropping areas to design a correct fertilizer management program. Inadequate supply of an essential element results in nutritional disorders that manifest themselves through alteration in the metabolic activities of the plant (MALAVOLTA, 2006).

The characteristics defined as deficiency symptoms and phosiological disorders, which can appear in leaves, stems or fruits. These disorders are related to the specific roles performed by each mineral element, whether macro or micronutrient (TAIZ; ZEIGER, 2013).

The uptake of nutrients occurs throughout the growth of the plant, increasing as the plant develops. The greatest demand for nutrients occurs in the period of flowering and the greatest development of the fruits, and in this same period, the plant becomes more vulnerable to the attack of pathogens, especially fungi and bacteria, a period in which it is possible to verify the greatest frequency of symptoms of mineral nutrient deficiency (FERREIRA et al., 2010).

The deficiency or excess of a mineral element, greatly influences the activity of others, and exerts marked effect, with consequences that have repercussions on the metabolism of the plant. It is also worth noting that the presence of an element in the soil does not necessarily imply its availability to the plant (TAIZ; ZIEGER, 2013).

The availability of nutrients can suffer interference from environmental conditions to soil type, such as pH, moisture and temperature, as well as in function of the amount of the element present in the soil, its form and solubility, and the assimilative capacity of the plant (HUBER, 1980).

The requirement for nitrogen (N) in tomato is greatest in the first stages of growth. N increases the dry matter weight of roots, stem, leaves and fruit, plant height, number of leaves, leaf area, flowering, fruit set and productivity. In its absence or insufficiency, plant growth is retarded and the older leaves become yellowish-green (FERREIRA et al., 2010).

Phosphorus (P) directly influences the growth rate of tomato plants, from the first stages of development. Due to its deficiency, the older leaves acquire a purplish color, due to the accumulation of the pigment anthocyanin. In later stages of development, the leaves show purple-brown areas that evolve to necrosis, falling prematurely and the plant delays its fruiting (MUELLER et al., 2015).

Potassium (K) is the most extracted nutrient in the solanaceae, due to the extensive flowering and fruiting period of the plant. It acts in the promotion of initial growth, in the accumulation of sugars, in the regulation of stomatic opening, is directly related to photosynthesis and consequently with the synthesis of photo-assimilates, enzyme activator and in the weight of the fruits. Potassium deficiency slows the growth of the plants, causing a reduction in the number and size of the fruits, also interfering with the firmness of the fruits, since the new leaves become thinner and the old ones present yellowing of the edges (KANAI et al., 2007).

Calcium (Ca) is a key component of the cells, maintaining the structure of the cell walls and stabilizing the cell membranes, it contributes to the structuring and rigidity of the cell wall of the fruit tissues, contributing to the increment of reproductive development of the plants. The process of pollen grain germination and pollen tube growth, which will ensure flower fertilization are also closely linked with Ca (MARSCHNER, 1995).

When the plant is not adequately supplied with Ca, lesions occur with the presence of depressed, dry and black necrosis, known as apical rot, as well as necrotic tissue inside the fruit, called black heart, and the death of the growth points may occur (EMBRAPA, 2003). Calcium is also a nutrient of great relevance in enzyme activation, in regulating the movement of water in cells and is essential for cell division, improving the firmness and size of the fruit, thus consequently productivity (MALAVOLTA, 2006).

Magnesium (Mg) is characterized by its activities in energy transfer and protein synthesis, as enzyme activators in respiration, photosynthesis and synthesis of DNA and RNA, improving flowering and plant production. The symptoms of deficiency are discoloration of the margins of the oldest leaflets, which progresses towards the interveinal area, with the nerves remaining green and later necrosing the tissues (MOGOR et al., 2013).

The nutrients K, Ca and Mg move to the aerial part through the transpiratory current present in the plant. Thus, deficiency symptoms will appear first in the oldest leaves (MALAVOLTA et al., 2006). Once incorporated into the cellular tissue, calcium is immobile, hence the need for a constant supply to meet the growth of the fruit (ALVARENGA, 2004).

Among the micronutrients, boron (B), which is involved in the production of nucleic acid and plant hormones, in the movement of sugars in plants and in the metabolism and translocation of carbohydrates, stand out. Zinc (Z) is correlated to the development and functioning of growth regulators produced by the plant, such as auxin, which influence the elongation of internodes, and may cause fruit splitting in cases of nutritional deficiency (RODRIGUES et al., 2002).

Iron (Fe) acts in reducing the production of chlorophyll in the plant, causing disturbances in the structure of chloroplasts, inhibiting the synthesis of lipids in plants, causing chlorosis of the leaves, which then become whitish. Manganese (Mn) is related to the production of chlorophyll in plants and with the greater rigidity of the cells, and can cause as deficiency symptoms light interveinal chlorosis in the old leaves, the spots can become purplish and the lower leaves totally yellow (OLIVEIRA et al., 2006).

In this context, foliar fertilization consists of supplying the plant with nutrients by spraying the aerial parts, making the mineral elements available in its leaves. Foliar application has the objective of correcting deficiencies of macro and micro nutrients, once symptoms have been visually detected in the leaves or identified through analysis of the leaf tissue in the laboratory (RODRIGUES et al., 2002).

Foliar application of nutrients is fast and efficient, surpassing soil fertilization, but it does not replace it, thus foliar fertilization is adopted as an additional strategy for the application of fertilizer, supplementing the crops throughout their development (BOARETTO et al., 1989).

References

ALBUQUERQUE, T. C. S. et al. Reguladores de crescimento vegetal na concentração de macronutrientes em videira itália. **Bragantia**, Campinas, v. 67, n. 3, p. 553-561, 2008.

ALVARENGA, M. A. R. Tomato: production in the field, greenhouse and hydroponics. Lavras: UFLA, 2004, p. 391.

ALMEIDA, G.; M. et al. Plant development through interference of auxins, cytokinins, ethylene and gibberylins. **Brazilian Journal of Applied Technology for Agricultural Science**, Guarapuava, v. 9, n. 3, p. 111-117, 2016.

ARYSTA LIVE SCIENCE DO BRASIL. Biozyme. Available at: https://www.upl-ltd.com/br/defensivos-agricolas/tratamento-de-sementes/biozyme. Accessed on: 02 jul. 2020.

BASTOS, A. R. R. et al. **Nutrição mineral e adubação do tomato**. 2. ed. Lavras: Editora Universitária. 2013. 130 p.

BINSFELD, J. A. et al. Use of bioactivator, biostimulant and nutrient complex in soybean seeds. **Pesquisa Agropecuária Tropical**, Brasília, v. 1, n. 1, p. 88-94, 2014.

BOARETTO, A. E. et al. **Foliar fertilization**. Campinas: Cargill Foundation, v. 2, 1989. 652 p.

BOURSCHEIDT, C. E. Bioestimulante e seus efeitos agronômicos na cultura da soja (*Glycine max*). **Pesquisa Agronômica Brasileira**, Ijuí, v. 28, p. 201-217, 2011.

DABADIA, A. C. A. Uso de bioestimulante na assimilação do nitrato e nos caracteres agronômicos em feijoeiro. **Cultura Agronômica**, Ilha Solteira, v. 24, n. 4, p. 321-332, 2015.

DANTAS, A. C. V. L. et al. Effect of gibberellic acid and the bioestimulant Stimulate® on the initial growth of thamarind. **Revista Brasileira de Fruticultura**, Jaboticabal, v. 34, n. 1, p. 08-14, 2012.

DOURADO NETO, D. et al. Application and influence of phytoregulator on corn plant growth. **Revista da Faculdade de Zootecnia, Veterinária e Agronomia**, São Carlos, v. 11, p. 93-102, 2004.

BRAZILIAN COMPANY OF AGRICULTURAL RESEARCH. EMBRAPA. **Cultivo de tomate para industrialização**. Brasília: EMBRAPA, v. 1, 2003. 13 p. (Technical Circular)

FAYAD, J. A. et al. Absorption of nutrients by tomato cultivated under field and protected environment conditions. Revista Horticultura Brasileira, Brasília, v. 20, p. 90-94, 2002.

FERREIRA, M. M. M. et al. Eficiência da adubação nitrogenada do tomatoiro em duas épocas de cultivo. **Revista Ceres**, Viçosa, v. 57, n. 2, p. 263-273, 2010.

FERREIRA, D. F. Sisvar: a computer statistical analysis system. **Ciência e Agrotecnologia**, Lavras, v. 35, n. 6, p. 1039-1042, 2011

GENUNCIO, G. C. et al. Production of tomato cultivars in hydroponics and fertigation under nitrogen and potassium ratios. **Horticultura Brasileira**, Brasília, v. 28, p. 446-452, 2010.

HORNKE, N. F. et al. Use of foliar fertilizer on the physiological performance of Moranga pumpkin seeds. **Tecnologia e Ciência Agropecuária**, São Paulo, v. 12, n. 3, p. 31-35, 2018.

HUBER, D. M. The role of mineral nutrition in defense. **Academic Press**, New York, v. 5, p.381-406,1980.

LIMA, A. A. et al. Foliar nutrient concentration and productivity of tomato cultivated under different substrates and doses of humic acids. **Revista Horticultura Brasileira**, Brasília, v. 29, p. 63-69, 2011.

MALAVOLTA, E. **Nutrição mineral de plantas**. São Paulo: Ed. Agr. Ceres. 2006. 631 p.

MARSCHNER, H. **Mineral nutrition of higher plants**. London: Academic Press, 1995.

MELO, L. D. F. A. et al. Action of phytoregulators in the culture of beans (*Phaseolus vulgaris L.*), sunflower (*Helianthus annuus L.)* and corn (*Zea mays* L.). **Educação Ambiental em Ação**, Rio Largo, v. 56, n. 15, p.1-5, 2016.

MOGOR, A. F. et al. Chlorophyll contents in tomato cultivars submitted to foliar applications of magnesium. **Pesquisa Agropecuária Tropical**, Goiânia, v. 43, n. 4, p. 363-369, 2013.

MUELLER, S. Modes of phosphorus application for two tomato cultivars. **Revista Horticultura Brasileira**, Brasília, v. 33, p. 356-361, 2015.

OLIVEIRA, R. H. Caracterização de sintomas visuais de deficiência de micronutrientes em tomatoiro do grupo salada. **Semina**, Londrina, v. 30, v. 11, p. 1093-1100, 2006.

PETRI, J. L et al. **Growth regulators for temperate climate fruit trees**. Epagri, 2016, 141 p.

RODRIGUES, D. S. et al. Amount absorbed and concentrations of micronutrients in tomato plants under protected cultivation. **Scientia Agricola**, Botucatu, v. 59, n. 1, p. 137-144, 2002.

TAIZ, L.; ZEIGER, E. Plant Physiology. 5.ed. Porto Alegre: Artmed, 2013. 954 p.

Chapter 2

FOLIAR FERTILIZER AND BIOSTIMULANT IN TOMATO PLANTS

Daiane Corrêa, Janiele Freire Polaciski and Suelen Cristina Uber

Introduction

Tomato (*Solanum lycopersicum* L.), belonging to the Solanaceae family is one of the most widespread vegetables in the world, occupying a prominent place on the consumer's table. The crop has adapted and develops practically in all geographical regions of the world, in different cropping systems and levels of cultural management, being widely spread in protected crops (MELO et al., 2016).

Tomato is present on the table of most of the population, being the second most produced and the first most consumed vegetable in the world, *in* various forms, from *fresh* consumption as a salad, to industrialized products, such as sauces and extracts. Its food and nutritional importance is characterized by the high concentration of lycopene, a powerful antioxidant, which increases immunity, helping the body protect against free radicals (SHAMI; MOREIRA, 2004).

The world tomato production was over 177 million tons, grown in more than 175 countries, an area of approximately 4.8 million hectares. Among the countries with the largest cultivation, China, India and the USA stand out. Brazil is the 9th largest producer of tomatoes in the world, with production exceeding 4.1 million tons, corresponding to 3% of the total fruit produced in the world (FAOSTAT, 2018).

Tomato cultivation stands out mainly in regions with a subtropical climate, in which the plants develop in environments with optimal temperatures between 18 and 32ºC. Cultivation in tropical regions has been growing gradually, especially with the establishment of agricultural areas for the production of tomatoes for industrial processing (BECKER et al. , 2016).

For the production of table tomatoes, destined for *fresh* consumption, especially in the Midwest region, characterized by the tropical climate, with high temperatures and rainfall rates, there is a lack of information on the vegetative and productive development of the tomato plant.

Biostimulants, which can be classified in the group of growth regulators or among foliar fertilizers, have been implemented in different agricultural crops, mainly in cereal grain and oilseed production, with the purpose of improving the development and productive performance of plants in agriculture (PETRI, et al. 2016).

Thus, foliar fertilizers and biostimulants emerge as an alternative to promote the cultivation of table tomatoes in the Midwest region, with potential use from seed treatment for seedling production to the flowering period of the plants, through foliar application.

In this context, the objective of this work was to evaluate the vegetative growth of the tomato plant as a function of foliar fertilizer and biostimulant application.

Material and Methods

The experiment was conducted at Mato Grosso State University (UNEMAT), University Campus of Alta Floresta-MT *campus* 2 (9°53'49.1" S 56°05'37.1" O) in the period from March to May 2020.

The plants were conducted during the period of germination, emergence and initial development in a greenhouse, with an average temperature of 26º C and daily irrigation. Subsequently, they were placed in a greenhouse, with ambient temperature between 22 and 38º C, and irrigated three times a week.

The climate of the region is type Am, according to the Koppen classification, being rainy tropical, reaching a high rainfall index in summer, which can reach values higher than 2,500 mm, and a dry winter, with high temperatures. The average annual temperature is around 25 °C (OLIVEIRA, 2006).

Before transplanting the seedlings into pots, soil analysis was performed, collected at a depth of 0-20 cm, to verify the need for correction and fertilization. The analysis was performed by the Laboratory of Soil, Fertilizer and Foliar Analysis - LASAF, UNEMAT. To determine the chemical and physical attributes of the soil the samples were submitted to analysis following the methodology described by EMBRAPA (SILVA, 2009).

The soil collected in the experimental area to fill the pots is classified as dystrophic Red Yellow Latosol (OLIVEIRA, 2006). The soil was corrected with limestone because its base saturation (V%) was below that recommended for the crop, which is 80% (RAIJ et al., 1996). Mineral fertilizers were also used, based on the results of soil analysis and technical recommendations and indications for tomato culture, with NPK corresponding to 50, 200 and 250 kg/ha (BECKER et al., 2016).

The experimental design was entirely randomized (DIC), with 5 treatments with increasing doses of biostimulant. For seedling production, four repetitions with 25 experimental units were adopted, totaling 100 seedlings per treatment. After transplanting to pots, with 10 L capacity, 6 replicates were used for each treatment, each consisting of one experimental unit (plant) per pot.

The treatments were composed of different doses of foliar fertilizer and biostimulant (Biozyme), being 2 ml/L, 4 ml/L, 6 ml/L, 8 ml/L and the control (0 ml/L), composed of only water. Each dose of biostimulant was measured using a syringe and inserted into a beaker containing 1 L of water.

Table 2 - Chemical composition with the water-soluble mineral elements, constituents of the foliar fertilizer and biostimulant used in the experiment.

Composition	% w/w	w/v (g/L)
Nitrogen	1,00	18
Potassium	5,00	60
Boron	0,08	0,96
Iron	0,40	4,8
Manganese	1,00	12
Sulfur	1,00	12
Zinc	2,00	24
Organic Carbon	3,50	42

Source: Arysta Life Science (2020).

The tomato seeds, cultivar Santa Clara 5.800, purchased commercially in the city of Alta Floresta-MT, were immersed in the solution composed of each treatment, containing the different doses for 3 minutes.

After this period, they were sown in 200-cell polyethylene trays, with 2 seeds allocated in each cell, which contained the commercial substrate Carolina Soil, composed of Sphagnum peat, expanded perlite, expanded vermiculite, roasted rice husk, and limestone. The thinning of the seedlings was performed 5 days after emergence (DAE).

The seedlings were transplanted at 20 DAE into pots, first placed in the greenhouse and later, at 30 DAE, taken to the greenhouse, covered with shingles on the sides and covered with transparent plastic film.

During the growth of the aerial part, three applications of biostimulants were made in each treatment, with different doses, the first at 15 DAE, the second at 27 DAE and the third at 40 DAE. Supplementation of all plants in the experiment with N, P, K, Ca and Mg was also performed through soil application at 30 days DAE, according to the indications and technical recommendations for the culture (BECKER et al., 2016).

The variables analyzed on the day of transplanting the plants from the trays to the pots, in the greenhouse, at 20 DAE, were plant height (PA) cm, root length (RC) cm, total plant length (TLC) cm, stem diameter (CD) cm, relative chlorophyll content (RCT) % and number of leaves (LE) n. in 10 plants, randomly selected in each treatment, the plants being discarded due to evaluations with destructive samples.

At 34 and 47 DAE, with the plants already in the pots and placed in the greenhouse, the variables AP, DC, NC, and CRT were evaluated in all plants of each treatment.

The height of the plant was measured from the height of the collar (cm) to its extremity, with the aid of a millimeter ruler.

Root length (cm) was estimated on the plants at 20 DAE, using a millimeter ruler, from the base of the collar to the tip of the largest root, after the roots were washed in water with constant agitation, to remove the substrate that was aggregated in the root system.

With the plant height and root length data, the total plant length (TPC) was estimated by summing the results obtained in the PA and CR evaluations at 20 DAE.

Stem diameter was expressed with the help of a digital caliper in (cm), which was inserted into the stem of the plant 3 cm above the soil surface.

The number of leaves was estimated by numerical counting (n.) of all the leaves present on the plant. The relative chlorophyll content (TRC) was determined by readings with a chlorophyllometer (%), Minolta SPAD-502 model, in the median part of the leaf, in 4 leaves per repetition.

The data were submitted to variance analysis and the means were compared using Tukey's test, at 5% probability, with the help of the statistical software SISVAR.

Results and Discussion

According to the variables analysed (Table 2), at 20 days after emergence, in function of the different doses of biostimulant, the 6 ml/L dose differed from the others, showing the best vegetative development, with 16.9 cm. However, the doses composed of the treatments 4 and 8 ml/L, equal to each other, also presented superior results when compared to the lowest dose of biostimulant tested and the control.

For root length, the 6 and 8ml/L doses were different from the others evaluated, however, they were equal among themselves, and obtained the greatest length of the root system.

Table 2- Plant height (PA) cm, root length (RC) cm, total plant length (TPC) cm, number of leaves (LE) n, stem diameter (CD) mm and relative chlorophyll content (RCT) of tomato (*Solanum lycopersicon* Mill.) plants at 20 days after emergence as a function of different doses of foliar fertilizer and biostimulant.

Treatments	AP	CR	CTP	NF	DC	TRC
Witness	13,5 c	10,2 b	23,3 c	4,0 b	0,3 a	29,7 b
2 ml/L	13,8 c	10,7 b	24,5 c	5,5 a	0,3 a	30,8 b
4 ml/L	15,1 b	11,0 b	26,1 b	5,8 a	0,4 a	30,9 b
6 ml/L	16,9 a	12,8 a	29,7 a	6,1 a	0,4 a	33,8 a
8 ml/L	15,2 b	12,5 a	27,7 b	5,9 a	0,4 a	33,4 a
CV %	6,9	10,5	7,3	7,9	4,2	8,6

Means followed by the same letter in the column do not differ, by Tukey's test, at 5% significance level.

According to Rodrigues et al. (2015), these biostimulant substances assist in greater absorption of water and nutrients, especially during the seedling production process, promoting increased root growth. The lowest doses of the biostimulant tested and the control did not show significant differences between them.

In the total length of the plant, the maximum dose showed the best growth results, demonstrating greater potential in the development of more vigorous seedlings due to the use of seed treatment with mineral elements and the application of biostimulant at 15 DAE.

According to Hornke et al (2018), higher doses of foliar fertilizer and biostimulant, provided better growth of pumpkin seedlings submitted to pumpkin seed treatment with Biozyme, in function of the product presenting macro and micronutrients, as well as hydrolyzed plant extracts in its composition, with high concentrations of cytokinin and zeatin, stimulating mainly root growth.

However, the intermediate doses, with 4 and 6 ml/L also promoted the development of tomato plants, when compared to those that were not supplemented with the biostimulant, highlighting the interaction of nutrients with foliar uptake, which can occur as efficiently as the supplementation of fertilizers through soil application.

According to Pereira (2002), the application of foliar fertilizers, due to the search for high yields and greater economy, the products are increasingly efficient in supplying nutritional requirements. Magalhães and Monnerat (1978), by foliar supplementation of plant nutrients, observed an increase in the production of dry mass in the roots and development of the aerial part of the plants, when the plants were subjected to fertilization.

For the analyzed variable of number of leaves, the control obtained the lowest leaf production when compared to the treatments in which the seeds were treated with different doses of biostimulant, and the others did not differ. The development of the aerial part can be increased in response to phytohormones linked to cell division and elongation, such as gibberellins (TAIZ; ZEIGER, 2013).

These results can be explained by the fact that foliar fertilizers and biostimulants work as activators of cell metabolism in the plant, give vigor to the immune system, reactivate physiological processes at different stages of development (PETRI et al., 2016), thus stimulating root growth, differentiation of axillary buds, inducing the formation of new shoots and leaves.

In relation to the diameter of the stem, there was no interaction between the treatments and doses evaluated, a fact that can be explained by the short period of development between the emergence and the evaluation, because the stems of vegetable plants usually present greater variation in relation to their development after transplanting the seedlings.

In the chlorophyll content we can observe that the doses 6m/L and 8m/L, stood out in relation to the other treatments evaluated, presenting higher chlorophyll content in the leaves. These results can be explained by the fact that the substances applied, present as functionality, the modification and increase of metabolic and physiological processes, through increased cell division and leaf elongation, as well as greater synthesis of chlorophyll, with action directly on photosynthesis, improving the uptake of nutrients and influencing productivity (SILVA et al., 2014).In watermelon seed treatment Radke et al. (2017), observed that foliar fertilizers and biostimulants with amino acids showed positive results in the initial development of seedlings, however the use of biostimulants can cause different results depending on the species used and the environmental conditions in which the plants are subjected, depending on temperature, precipitation and relative humidity.For the plants evaluated at 34 DAE, according to the results presented (Table 3), the higher doses of biostimulants provided greater development of the aerial part of the plants. It should be noted that in this period, the plants had already undergone a second foliar application with the use of biostimulant, favoring the aerial development of the tomato plants.

Table 3- Plant height (PA) cm, stem diameter (CD) cm, number of leaves (LE) n and relative chlorophyll content (RCT) of tomato (*Solanum lycopersicon* Mill.) plants at 34 days after emergence as a function of different doses of foliar fertilizer and biostimulant.

Treatments	AP	DC	NF	TRC
Witness	18,9 b	0,6 a	6,9 c	36,2 b
2 ml/L	19,7 b	0,7 a	8,2 b	38,8 a
4 ml/L	21,4 a	0,7 a	8,5 b	38,8 a
6 ml/L	21,8 a	0,9 a	10,7 a	39,6 a
8 ml/L	21,6 a	0,8 a	10,1 a	39,1 a
CV %	8,3	5,7	9,5	11,9

Means followed by the same letter in the column do not differ, by Tukey's test, at 5% significance level.

According to the results presented, for the diameter of the stem there were no significant differences among the treatments tested, with variations between 0.6 and 0.9 cm. For the variable number of leaves, there was a greater amount of leaf area in the doses 6 and 8m/L as a function of the increased number of leaves when compared to the control treatment. Considering the greater height of the plant, it can be estimated that these plants also have greater fresh mass of the aerial part, consequently with more leaves.

For the relative chlorophyll content, all the doses of foliar fertilizers and biostimulants were higher than the control, with higher levels of chlorophyll. The greater quantity of chlorophyll in the plants may indicate the greater quantity of nitrogen accumulation in the tissue of the cells, contributing to the reduction of evapotranspiration, potentiating the development of the plant (ARGENTA et al., 2001).

According to the results observed, it is possible to verify the dose response effect, in which as the dose used increases, it provides an increase in the growth of plant height, due to the higher concentrations of fertilizers used in the supplementation of tomato plants.

In this way, the cause and dependence effect can still be correlated, with the higher concentration of chlorophyll in the treatments with the highest doses of the foliar fertilizer and biostimulant, a factor that is a function of the greater availability of nutrients, such as N, which favor the absorption of light energy and the accumulation of photo-assimilates in leaves with higher chlorophyll levels (VENDRUSCULO et al., 2017).

According to Petri et al. (2016) foliar fertilizers and biostimulants can increase the aerial development of plants, production and resistance to stresses caused by temperature and water deficit, correlated with the higher chlorophyll content present in the leaves. Compared to Silva (2014), for the variables plant height, number of leaves, stem diameter, length of the largest root, and root volume, there was interaction between the biostimulants and the factors evaluated.

In this context, it can be observed that the use of biostimulant-based products, containing macro and micronutrients, can cause an increase in plant growth, both in root length, aerial part, and plant development.

Conclusion

Through this work, it can be verified that the use of foliar fertilizer and biostimulant promotes greater vegetative development of tomato plants.

The doses of 6 and 8 ml/L of foliar fertilizer and biostimulant provided the highest growth of the aboveground part, root system, chlorophyll accumulation, and number of leaves in tomato plants.

References

ALBUQUERQUE, T. C. S. et al. Reguladores de crescimento vegetal na concentração de macronutrientes em videira itália. **Bragantia**, Campinas, v. 67, n. 3, p. 553-561, 2008.

ALVARENGA, M. A. R. Tomato: production in the field, greenhouse and hydroponics. Lavras: UFLA, 2004, p. 391.

ALMEIDA, G.; M. et al. Plant development through interference of auxins, cytokinins, ethylene and gibberylins. **Brazilian Journal of Applied Technology for Agricultural Science**, Guarapuava, v. 9, n. 3, p. 111-117, 2016.

AQUINO, H. F. **Morphological, agronomic characterization and genetic divergence of pepper accessions.** 2016. 91 f. Dissertation (Master in Agronomy), Universidade Federal Rural de Pernambuco, Recife, 2016.

ARYSTA LIVE SCIENCE DO BRASIL. Biozyme. Available at: https://www.upl-ltd.com/br/defensivos-agricolas/tratamento-de-sementes/biozyme. Accessed on: 02 jul. 2020.

BASTOS, A. R. R. et al. **Nutrição mineral e adubação do tomato.** 2. ed. Lavras: Editora Universitária. 2013. 130 p.

BECKER. W. F. et al. **Integrated production system for tutored tomato.** Florianópolis: Epagri, 2016. 149 p.

BINSFELD, J. A. et al. Use of bioactivator, biostimulant and nutrient complex in soybean seeds. **Pesquisa Agropecuária Tropical**, Brasília, v. 1, n. 1, p. 88-94, 2014.

BOARETTO, A. E. et al. **Foliar fertilization.** Campinas: Cargill Foundation, v. 2, 1989. 652 p.

BOURSCHEIDT, C. E. Bioestimulante e seus efeitos agronômicos na cultura da soja (*Glycine max*). **Pesquisa Agronômica Brasileira**, Ijuí, v. 28, p. 201-217, 2011.

NATIONAL SUPPLY COMPANY. CONAB. Compêndio de Estudos Conab, v. 1. Brasília: Conab, 2019. 22 p.

CARVALHO, N. M.; NAKAGAWA, J. **Sementes: ciência, tecnologia e produção.** 4ª ed. Jaboticabal: Funep. 2000. 588 p.

DABADIA, A. C. A. Uso de bioestimulante na assimilação do nitrato e nos caracteres agronômicos em feijoeiro. **Cultura Agronômica**, Ilha Solteira, v. 24, n. 4, p. 321-332, 2015.

DANTAS, A. C. V. L. et al. Effect of gibberellic acid and the bioestimulant Stimulate® on the initial growth of thamarind. **Revista Brasileira de Fruticultura**, Jaboticabal, v. 34, n. 1, p. 08-14, 2012.

DOSSA, D.; FUCHS, F. Tomato: technical-economic analysis and the main indicators of production in world, Brazilian and Paraná markets. Boletim Técnico 03 Tomate, Curitiba, Aug. 2017. Available at: http://www.ceasa.pr.gov.br/arquivos/File/BOLETIM/Boletim_Tecnico_Tomate1.pdf. Accessed on: 05 Aug. 2020.

DOURADO NETO, D. et al. Application and influence of phytoregulator on corn plant growth. **Revista da Faculdade de Zootecnia, Veterinária e Agronomia**, São Carlos, v. 11, p. 93-102, 2004.

BRAZILIAN COMPANY OF AGRICULTURAL RESEARCH. EMBRAPA. **Cultivo de tomate para industrialização**. Brasília: EMBRAPA, v. 1, 2003. 13 p. (Technical Circular)

BRAZILIAN COMPANY OF AGRICULTURAL RESEARCH. EMBRAPA. **Cultivo de Tomate.** Brasília: EMBRAPA, v. 3, 2010. 28 p. (Technical Circular)

FAOSTAT. FAO. Tomato Production. 2018. Available at: http://www.fao.org/faostat/en/#home. Accessed on: 10 Aug. 2020.

FAYAD, J. A. et al. Absorption of nutrients by tomato cultivated under field and protected environment conditions. Revista Horticultura Brasileira, Brasília, v. 20, p. 90-94, 2002.

FILGUEIRA, F. A. R. **Novo Manual de Olericultura: agrotecnologia moderna na produção e comercialização de hortaliças**. Lavras: Editora UFLA, 2008. 333 p.

FERREIRA, A. G., BORGHETTI, F. **Germination: do básico ao aplicado**. Porto Alegre: Artimed. 2004. 323 p.

FERREIRA, M. M. M. et al. Eficiência da adubação nitrogenada do tomatoiro em duas épocas de cultivo. **Revista Ceres**, Viçosa, v. 57, n. 2, p. 263-273, 2010.

FERREIRA, D. F. Sisvar: a computer statistical analysis system. **Ciência e Agrotecnologia**, Lavras, v. 35, n. 6, p. 1039-1042, 2011

GENUNCIO, G. C. et al. Production of tomato cultivars in hydroponics and fertigation under nitrogen and potassium ratios. **Horticultura Brasileira**, Brasília, v. 28, p. 446-452, 2010.

GUIMARÃES, M. A. et al. Productivity and fruit flavor of salad group tomatoes as a function of pruning. **Bioscience Journal**, Uberlândia, v. 24, n. 1, p. 32-38, 2008.

HARTMANN, H. T. et al. **Plant propagation: principles and practices**. 7. ed. New Jersey: Prentice Hall, 2002. 880 p.

HORNKE, N. F. et al. Use of foliar fertilizer on the physiological performance of Moranga pumpkin seeds. **Tecnologia e Ciência Agropecuária**, São Paulo, v. 12, n. 3, p. 31-35, 2018.

HUBER, D. M. The role of mineral nutrition in defense. **Academic Press**, New York, v. 5, p.381-406,1980.

BRAZILIAN INSTITUTE OF GEOGRAPHY AND STATISTICS. IBGE. Brazilian Agricultural Production 2019. Brasília: IBGE, 2019. Available at: https://sidra.ibge.gov.br/tabela/5457. Accessed on: 18 Jul. 2020.

KANAI, S. et al. Depression of sink activity precedes the inhibition of biomass production in tomato plants subjected to potassium deficiency stress. **Journal of Experimental Botany**, Amsterdan, v. 58, p. 2917-2928, 2007.

LIMA, A. A. et al. Foliar nutrient concentration and productivity of tomato cultivated under different substrates and doses of humic acids. **Revista Horticultura Brasileira**, Brasília, v. 29, p. 63-69, 2011.

MATOS, E. S.; SHIRAHIGE, F. H.; MELO, P. C. T. Desempenho de híbridos de tomato de crescimento indeterminado em função de sistemas de condução de planta. **Revista Horticultura Brasileira**, Brasília, v. 30, n. 2, p. 240-245, 2012.

MACHADO, A. Q. et al. Produção de tomate italiano (saladete) sob diferentes densidades de plantio e sistemas de prune visando ao consumo *in natura*. **Revista Horticultura Brasileira**, Brasília, v. 25, p. 149-153, 2007.

MAGALHÃES, J. R.; MONNERAT, P. H. Aplicação foliar de boron na prevenção de deficiência e na composição mineral do tomatoiro. **Pesquisa Agropecuária Brasileira**, Brasília, v. 13, p. 73-80, 1978

MALAVOLTA, E. **Nutrição mineral de plantas**. São Paulo: Ed. Agr. Ceres. 2006. 631 p.

MARSCHNER, H. **Mineral nutrition of higher plants**. London: Academic Press, 1995.

MARTINS, G. Cultivo em ambiente protegido. Viçosa: UFV, 2000. 148 p.

MELO, P.C.T. et al. Performance of tomato cultivars in organic system under protected cultivation. **Revista Horticultura Brasileira**, Brasília, v. 27, p. 553-559, 2009.

MELLO, S. C.; C. VITTI, G. C. Tomatoiro development and modifications in soil chemical properties as a function of organic residue application under protected cultivation. **Revista de Horticultura Brasileira**, Brasília, v. 20, n. 2, p. 275-284, 2009.

MELO, L. D. F. A. et al. Action of phytoregulators in the culture of beans (*Phaseolus vulgaris* L.), sunflower (*Helianthus annuus L.)* and corn (*Zea mays* L.). **Educação Ambiental em Ação**, Rio Largo, v. 56, n. 15, p.1-5, 2016.

MELO, A. et al. Solanaceae in organic system in Brazil: tomato, potato and physalis. **Scientia Agropecuária**, Campinas, v. 8, n. 3, p. 279-290, 2017.

MOGOR, A. F. et al. Chlorophyll contents in tomato cultivars submitted to foliar applications of magnesium. **Pesquisa Agropecuária Tropical**, Goiânia, v. 43, n. 4, p. 363-369, 2013.

MORITZ, B.; TRAMONTE, V. L. C. Lycopene bioavailability. **Revista Nutrição**, Campinas, v. 19, n. 2, p. 265-273, 2006.

MUELLER, S. Modes of phosphorus application for two tomato cultivars. **Revista Horticultura Brasileira**, Brasília, v. 33, p. 356-361, 2015.

NAIKA, S. **A cultura do tomato**: produção, processamento e comercialização. v. 1, 2005. 105 p.

NAVARRETE, M.; JEANNEQUIN, B. Effect of frequency of axillary bud pruning on vegetative growth and fruit yield greenhouse tomato crops. **Scientia Horticulturae**, Amsterdam, v. 86, n. 3, p.197-210, 2000.

OLIVEIRA, R. H. Caracterização de sintomas visuais de deficiência de micronutrientes em tomatoiro do grupo salada. **Semina**, Londrina, v. 30, v. 11, p. 1093-1100, 2006.

PERALTA, I. E. et al. Taxonomy of wild tomatoes and their relatives (*Solanum* lycopersicon-Solanaceae). **Systematic Botany Monographs**, Cuyo, v. 84, n. 2, 2008.

PETRI, J. L et al. **Growth regulators for temperate climate fruit trees**. Epagri, 2016, 141 p.

RADKE, A. K. et al. Amino acids via seed treatment: reflections on watermelon seed vigor. **Tecnologia & Ciência Agropecuária**, Brasília, v. 11, p. 113-117, 2017.

REIS, L. S. et al. Evapotranspiration and crop coefficient of kaki tomato grown in a protected environment. **Revista Brasileira de Engenharia Agrícola e Ambiental**, Campina Grande, v. 13, p. 289-296, 2009.

RODRIGUES, D. S. et al. Amount absorbed and concentrations of micronutrients in tomato plants under protected cultivation. **Scientia Agricola**, Botucatu, v. 59, n. 1, p. 137-144, 2002.

SHAMI, N. J. I. E.; MOREIRA, E. A. M. Licopene as an antioxidant agent. **Revista Nutrição**, Campinas, v. 17, n. 2, 2004

SILVA, J. V. Aproveitamento de materiais alternativos na produção de mudas de tomateiro sob adubação foliar. **Revista Ciência Agronômica**, Campinas, v. 45, n. 3, p. 528-536, 2014.

SILVA, F. C. **Manual de análises químicas de solos, plantas e fertilizantes**. 2. ed. Brasília: Embrapa Informação Tecnológica, 2009. 627 p.

TAIZ, L.; ZEIGER, E. Plant Physiology. 5.ed. Porto Alegre: Artmed, 2013. 954 p.

VENDRUSCOLO, E. P. et al. Physicochemical changes in fruit of lacy melon under application of biostimulant. **Revista Colombiana de Ciencias hortícolas**, Tunja Boyacá, v. 11, n. 2, p. 459-463, 2017.

Chapter 3

FOLIAR FERTILIZER AND BIOSTIMULANT IN PEPPER PLANTS

Daiane Corrêa, Camila Penteado Zenaro and Suelen Cristina Uber

Introduction

The Italian sweet pepper (*Capsicum annuum*), which belongs to the Solanaceae family, presents a good regional appreciation, being a plant of easy cultivation, which can be consumed in several ways, presenting different sizes, shapes, colors and with different levels of pungency (MESQUITA et al., 2016). They are foods of quite diversified use that can be present every day on the Brazilian table, because it can be found canned or *in natura*, it is widely consumed in typical foods of each region, and can also be used in dishes with decorative purposes, besides being considered a great plant with medicinal and pharmacological principles.

In Brazil, the production of chili peppers has been gaining strength in the last years, moving the market in more than 100 million reais per year, being cultivated mainly in regions with subtropical climate, as well as it has adapted to warmer environments, such as tropical climate, obtaining great productive characteristics. Among the Brazilian states that stand out as the largest producers are Minas Gerais, Goiás, São Paulo, Rio Grande do Sul and Ceará (BARBOSA et al., 2002).

Besides, chili cultivation is of great socioeconomic importance, due to its income generation and social development, especially among family farmers, who are responsible for most of its cultivation. The country has an area of 5 thousand hectares of production, and its productivity presents variations depending on the cultivar used, between 10 and 30 tons per hectare (VIEIRA; CASTRO, 2004).

The pepper plant has been gaining appreciation in the agribusiness mainly because they are cultivated in almost all Brazilian regions, thus the production of seedlings is a crucial phase to have a satisfactory production. The cultivation in tropical regions has been gradually growing, especially with the implementation of cultivation areas in protected environments. However, there is a lack of information about the development of pepper plants under these conditions.

The physiological quality of seeds is one of the main factors that result in the expressive potential of germination and emergence, quality and productivity of almost all crops. Among the factors that interfere with seed germination, limiting the emergence and initial development of seedlings is the temperature, either too low or too high during the germination process (AMARO et al., 2015).

There are foliar fertilizers and biostimulants, as well as plant regulators that can be used to stimulate parthenocarpy and contribute to improving seedling development, aiming to obtain better uniformity of plants for transplanting, such as gibberellins, auxins, cytokinins, and mineral compounds (CANO-MEDRANO; DARNELL, 1997).

Foliar fertilizers and biostimulants, which can be classified in the group of growth regulators or among foliar fertilizers, have been implemented in different agricultural crops, both in seed treatment and in aerial part sprays, through the addition of different compounds with macro and micronutrients, in which some studies show efficiency by presenting action similar to those of plant hormones that are already present in the market (COSTA et al., 2017), with the purpose of improving plant development in agriculture (PETRI et al., 2016).

Thus, foliar fertilizers and biostimulants emerge as an alternative to promote the cultivation of pepper in regions with higher temperatures, with potential use from seed treatment for the production of seedlings until the flowering period of the plants, through foliar application.

In this context, the objective of this work was to evaluate the effect of treating Italian sweet pepper seeds with foliar fertilizer and biostimulant on seedling production.

Material and Methods

The experiment was conducted at Mato Grosso State University (UNEMAT), University Campus of Alta Floresta-MT *campus* 2 (9°53'49.1" S 56°05'37.1" O) in the period from November to December 2020.

The plants were conducted during the period of germination, emergence, and initial development in a greenhouse with an average temperature of 28º C and daily irrigation.

The climate of the region is type Am, according to the Koppen classification, being rainy tropical, reaching a high rainfall index in summer, which can reach values higher than 2,500 mm, and a dry winter, with high temperatures. The average annual temperature is around 25 °C (OLIVEIRA, 2006).

The experimental design was entirely randomized (DIC), with 5 treatments with increasing doses of foliar fertilizer and biostimulant. For seedling production, five repetitions with 10 experimental units were adopted, totaling 50 seedlings per treatment.

The treatments were composed of different doses of foliar fertilizer and biostimulant (Biozyme), being 2 ml, 4 ml, 8 ml, 16 ml and the control (0 ml), composed of water only.

Each dose of biostimulant was measured using a syringe and inserted into a beaker containing 200 mL of water.

Table 1 - Chemical composition with the water-soluble mineral elements, constituents of the foliar fertilizer and biostimulant used in the experiment.

Composition	% w/w	w/v (g/L)
Nitrogen	1,00	18
Potassium	5,00	60
Boron	0,08	0,96
Iron	0,40	4,8
Manganese	1,00	12
Sulfur	1,00	12
Zinc	2,00	24
Organic Carbon	3,50	42

Source: Arysta Life Science (2020).

The sweet pepper seeds, Italian cultivar, purchased commercially in the city of Alta Floresta - MT, were immersed in the solution composed of each treatment, containing the different doses for 3 minutes.

After this period, they were sown in 200-cell polyethylene trays, with 2 seeds allocated in each cell, which contained the commercial substrate Carolina Soil, composed of Sphagnum peat, expanded perlite, expanded vermiculite, roasted rice husk, and limestone. The thinning of the seedlings was performed 5 days after emergence (DAE).

The variables analyzed at 20 and 35 DAE, were plant height (PA) cm, root length (RC) cm, total plant length (CTP) cm, stem diameter (CD) cm, relative chlorophyll content (TRC) % and number of leaves (NF) n. in 10 plants, randomly chosen in each treatment, being the plants discarded due to evaluations with destructive samples.

The plant height was measured by estimating the height of the collar (cm) up to its upper extremity, using a millimeter ruler.

Root length (cm) was estimated on the plants with the aid of a millimeter ruler, from the base of the collar to the tip of the largest root, after the roots were washed in water with constant agitation, to remove the substrate that was aggregated in the root system.

With the plant height and root length data, the total plant length (TPC) was estimated, through the sum of the results obtained in the PA and CR evaluations.

Stem diameter was expressed using a digital caliper in (mm), which was inserted into the plant stem 5 mm above the surface of the substrate.

The number of leaves was estimated by numerical counting (n.) of all the leaves present on the plant. The relative chlorophyll content (TRC) was determined by readings with a chlorophyllometer (%), Minolta SPAD-502 model, in the median part of the leaf, in 10 leaves per repetition.

The data obtained from the variables were submitted to variance analysis and the means were compared using the Tukey test, at 5% probability, with the help of the SAS statistical software.

Results and Discussion

For the variables of plant height, root growth and total plant growth, it can be observed that the best development results were obtained from the application of different doses of foliar fertilizer and biostimulant (Table 2). However, the doses that were applied did not show significant differences between them, differing only from the control. This fact can be explained by the fact that the nutrients present in the foliar fertilizer and biostimulant contribute to the best initial development of the plant, increasing the results obtained.

Table 2- Plant height (PA) cm, root length (RC) cm, total plant length (TPC) cm, stem diameter (CD) mm, number of leaves (LE) n and relative chlorophyll content (RCT) of pepper plants (*Capsicum annuum)* at 20 days after emergence as a function of different doses of foliar fertilizer and biostimulant.

Treatments	AP	CR	CTP	DC	NF	TRC
Witness	3,0 b	3,5 b	6,5 b	1,0 a	2,0 a	27,0 a
2 ml/L	5,1 a	5,4 a	10,5 a	1,0 a	2,0 a	30,4 a
4 ml/L	5,9 a	6,2 a	12,1 a	1,0 a	2,1 a	31,1 a
8 ml/L	5,6 a	5,9 a	11,5 a	1,0 a	2,1 a	30,7 a
16 ml/L	4,8 a	5,6 a	10,4 a	1,0 a	2,1 a	30,5 a
CV %	10,8	13,7	14,1	10,3	9,8	12,9

Means followed by the same letter in the column do not differ, by Tukey's test, at 5% significance level.

Hornke et al. (2018), can observe that the higher doses of biostimulant, can provide growth of pumpkin seedlings submitted to the treatment of pumpkin seeds with Biozyme. Where macro and micronutrients and the hydrolyzed plant extracts in its composition, with high concentrations of cytokinin and zeatin, mainly stimulated root growth, linked to increased development of the aerial part of the plant.

Polacinski (2020), verified that the doses 4 and 8 ml/L of foliar fertilizer and biostimulant, applied to tomato (*Solanum lycopersicum* L.), did not show significant differences when compared to each other; however, they presented superior results to the other doses applied in his study.

According to Rodrigues et al. (2015), foliar fertilizers and biostimulants contribute to greater water and nutrient uptake for plants, especially during the seedling production process, promoting increased root growth.

Radke et al. (2017), treating watermelon seeds observed that biostimulants plus amino acids showed positive results in the initial development of seedlings. However, the use of foliar fertilizers and biostimulants can cause different results depending on the species used and the environmental conditions in which the plants are subjected, depending on temperature, precipitation, and relative humidity.

For the variable stem diameter (CD), number of leaves (LE) and relative chlorophyll content (RCT), the applied doses did not show significant differences between the treatments. Because the plants were still at 20 DAE, in other words, very young, as the days pass they may present increased growth and consequently improve the relative indices of their stem diameter and number of leaves. With the high initial growth rate of the plants and because most of the plant's leaf area consists of young leaves, with longer periods of development, containing high photosynthetic capacity, the TRC has the potential to increase its production (AUMONDE et al., 2011). Correlated to this, the chlorophyll content of the plant with the passing of the days will have the greatest accumulation of photo-assimilates, contributing to the increase in the number of leaves and stem diameter. Polacinski (2020), describes that for the variable of stem diameter, there was no increase in development, according to the results presented between the treatments with the different doses evaluated, a fact that can be justified by the short period of development between emergence and the completion of the evaluation. We can observe that in Table 3, for the variables plant height (PA), root length (RC), and total plant length (CTP), the results were the same, but the control treatment obtained the lowest measurements in relation to the other treatments.

Table 3- Plant height (PA) cm, root length (RC) cm, total plant length (TPC) cm, stem diameter (CD) mm, number of leaves (LE) n and relative chlorophyll content (RCT) of pepper plants (*Capsicum annuum*) at 35 days after emergence as a function of different doses of foliar fertilizer and biostimulant.

Treatments	AP	CR	CTP	DC	NF	TRC
Witness	6,3 b	5,9 b	12,2 b	2,0 a	6,3 a	29,3 b
2 ml/L	8,4 a	7,8 a	16,2 a	2,3 a	6,3 a	32,1 a
4 ml/L	9,3 a	8,4 a	17,7 a	2,6 a	6,4 a	32,6 a
8 ml/L	8,5 a	8,7 a	17,2 a	2,4 a	6,5 a	32,5 a
16 ml/L	6,9 b	6,4 b	13,3 b	2,2 a	7,0 a	32,9 a
CV %	14,5	12,1	13,9	7,3	11,8	9,4

Means followed by the same letter in the column do not differ, by Tukey's test, at 5% significance level.

In the dose of 16 ml/L the averages were lower than the other doses with the application of foliar fertilizer and biostimulant, and it can be observed that this dose can behave as a limiting dose to the absorption capacity of the plant, since the plant has already absorbed what is necessary to maintain itself and does not use or store the rest, thus, it can contribute to interfere negatively in the development of the plant, both in its root growth and in its diameter.

The variables stem diameter (CD), number of leaves (LE) and relative chlorophyll content, showed no significant differences between them, even at 35 DAP the plants showed no improvement in development when compared to 20 DAP. However, the difference is present at 35 DAP when we observe the relevance when biofertilizer is applied, where there is a slight increase in plant development.

Corroborating the results obtained in this study Argenta et al. (2001), obtained the same result by observing that for the relative chlorophyll contents for all doses of foliar fertilizers and biostimulants, the medians were higher than the control.

These results can be explained by the fact that the substances applied to the plant have functions that contribute to the modification and increase of metabolic and physiological processes, through increased cell division and leaf elongation, as well as greater synthesis of chlorophyll, with action directly on photosynthesis, improving the absorption of nutrients and influencing the productivity and development of the plant (SILVA et al., 2014).

According to Pereira (2002), with the application of foliar fertilizers, seeking high yields and greater economy, the products are increasingly efficient in supplying the nutritional requirements of the plants, in order to obtain better yields.

Thus, the results acquired with the use of biofertilizers are due to their functioning, contributing as enzyme activators of secondary metabolites of plants, giving vigor and acting at various stages of development (PETRI et al., 2016), thus stimulating root growth, stem diameter, number of leaves and chlorophyll content of plants, increasing their productive and economic yield for the producer.

Conclusion

However, it can be concluded that the treatment of Italian sweet pepper seeds with foliar fertilizer and biostimulant, presents beneficial effects, improving plant development during seedling production.

The doses of 2, 4 and 8 ml/L of foliar fertilizer and biostimulant showed the best response in the growth of pepper seedlings.

References

ABREU, L. F. et al. Effect of drying on the antioxidant properties of *Capsicum annuum* var. annuum red peppers. In: Embrapa Amazônia Oriental-Article in annals of congress (ALICE). In: CONGRESSO BRASILEIRO DE CIÊNCIA E TECNOLOGIA DE ALIMENTOS, 25., 2016, Gramado. **Annals...** Gramado: Regional SBCTA, 2016.

ABUD, H. F. et al. Physiological quality of seeds of malagueta and biquinho peppers during ontogenesis. **Pesquisa Agropecuária Brasileira**, Brasília, v. 48, n. 12, p. 1546-1553, 2013.

ALBUQUERQUE, T. C. S. et al. Reguladores de crescimento vegetal na concentração de macronutrientes em videira itália. **Bragantia**, Campinas, v. 67, n. 3, p. 553-561, 2008.

ALMEIDA, G.; M. et al. Plant development through interference of auxins, cytokinins, ethylene and gibberylins. **Brazilian Journal of Applied Technology for Agricultural Science**, Guarapuava, v. 9, n. 3, p. 111-117, 2016.

AMARO, G. B. et al. **Knowledge tree:** pepper. 2012. Available at: <http://www.agencia.cnptia.embrapa.br/gestor/pimenta/ Opening.html>. Accessed on: 1 NOV. 2020.

ARGENTA, G. et al. Relation of chlorophyllometer reading with extractable chlorophyll and nitrogen contents in maize leaf. **Revista Brasileira de Fisiologia Vegetal**, v. 13, n. 2, p. 158-167, 2001.

AUMONDE, T. Z. et al. Growth analysis of grafted and non-grafted mini watermelon hybrid Smile®. **Interciencia**, Rio Verde, v. 36, n. 9, p. 677-681, 2011.

BARBOSA, R. I. Pimentas do gênero capsicum cultivadas em Roraima, Amazônia Brasileira. Domesticated species. **Acta Amazônica**, Belém, v. 3, n. 2, p. 177-132, 2002.

BIANCHETTI, L. B. Aspectos morfológicos, ecológicos e biogeográficos de 10 táxons de *Capsicum* (Solanaceae) ocorrentes no Brasil. **Acta Botânica Brasílica**, Brasília, v. 10, n. 2, p. 393-394, 1996.

BINSFELD, J. A. et al. Use of bioactivator, biostimulant and nutrient complex in soybean seeds. **Pesquisa Agropecuária Tropical**, Brasília, v. 1, n. 1, p. 88-94, 2014.

BOURSCHEIDT, C. E. Bioestimulante e seus efeitos agronômicos na cultura da soja (*Glycine max*). **Pesquisa Agronômica Brasileira**, Ijuí, v. 28, p. 201-217, 2011.

BOSLAND, P. W.; VOTAVA, E. J. **Peppers: vegetable and spice *Capsicums*.** 2. ed. Cambridge: CABI - Crop Production Science in Horticulture, 2012. 230p. (Série 22).

CARVALHO, S. I. C.; BIANCHETTI L.; BUSTAMANTE PG; SILVA DB. Catálogo **de germoplasma de pimentas e pimentões (*Capsicum* spp) da Embrapa Hortaliças.** Brasília: Embrapa Hortaliças, 2003. 49 p. (Embrapa Hortaliças. Documents, 49).

NATIONAL SUPPLY COMPANY. CONAB. Compêndio de Estudos Conab, v. 1. Brasília: Conab, 2019. 22 p.

COSTA, J. P. B. M. Quality of pepper seedlings produced in vermiculite and submitted to salt stress and biostimulant. Conference: IV Inovagri International Meeting, **Proceedings...** Amsterda, 2017.

DABADIA, A. C. A. Uso de bioestimulante na assimilação do nitrato e nos caracteres agronômicos em feijoeiro. **Cultura Agronômica**, Ilha Solteira, v. 24, n. 4, p. 321-332, 2015.

DANTAS, A. C. V. L. et al. Effect of gibberellic acid and the bioestimulant Stimulate® on the initial growth of thamarind. **Revista Brasileira de Fruticultura**, Jaboticabal, v. 34, n. 1, p. 08-14, 2012.

DOURADO NETO, D. et al. Application and influence of phytoregulator on corn plant growth. **Revista da Faculdade de Zootecnia, Veterinária e Agronomia**, São Carlos, v. 11, p. 93-102, 2004.

BRAZILIAN COMPANY OF AGRICULTURAL RESEARCH. EMBRAPA. **Cultivo de tomate para industrialização**. Brasília: EMBRAPA, v. 1, 2003. 13 p. (Technical Circular)

BRAZILIAN COMPANY OF AGRICULTURAL RESEARCH. EMBRAPA. **Cultivo de Tomate.** Brasília: EMBRAPA, v. 3, 2010. 28 p. (Technical Circular)

FOROUHI, N. G. Consumption of hot spicy foods and mortality: is chilli good for your health? **Journal Science,** Laedgaaard, v. 351, p. 41-48, 2015.

GENUNCIO, G. C. et al. Production of tomato cultivars in hydroponics and fertigation under nitrogen and potassium ratios. **Horticultura Brasileira**, Brasília, v. 28, p. 446-452, 2010.

HOEHNE, F. C.; Botânica e agricultura no Brasil no século XVI: pesquisas e contribuições. **Brasiliana**, 1937. 98 p.

HORNKE, N. F. et al. Use of foliar fertilizer on the physiological performance of Moranga pumpkin seeds. **Tecnologia e Ciência Agropecuária**, São Paulo, v. 12, n. 3, p. 31-35, 2018.

BRAZILIAN INSTITUTE OF GEOGRAPHY AND STATISTICS. IBGE. Brazilian Agricultural Production 2019. Brasília: IBGE, 2019. Available at: https://sidra.ibge.gov.br/tabela/5457. Accessed on: 18 Jul. 2020.

LIMA, A. A. et al. Foliar nutrient concentration and productivity of tomato cultivated under different substrates and doses of humic acids. **Revista Horticultura Brasileira**, Brasília, v. 29, p. 63-69, 2011.

LOPES, C. A. et al.; **Pimenta (*Capsicum* spp.).** 2007. EMBRAPA. Available at: <https://sistemasdeproducao.cnptia.embrapa.br/FontesHTML/Pimenta/Pimenta_capsicum _spp/index.html>. Accessed on: 01 Nov. 2020.

MAZUHOVITZ, S. C.; C. VITTI, G. C. Tomatoiro development and modifications in soil chemical properties as a function of organic waste application under protected cultivation. **Revista de Horticultura Brasileira**, Brasília, v. 20, n. 2, p. 275-284, 2009.

MELO, A. M. T.; NASCIMENTO, W. M.; FREITAS, R. A. Produção de sementes de pimenta. In: NASCIMENTO, W. M. **Produção de sementes de hortaliças**. Brasília: EMBRAPA, 2014. p. 169-197.

MOREIRA, G. R. et al. Espécies e variedades de pimenta. **Informe Agrocuário**, São Paulo, v. 27, p. 16-29. 2006.

MOREIRA, B. V.; LEITE, M. S. New cultivar of Cumari pepper: characterization of a strain and application for cultivar protection. 2014. 40 f. >Monograph (Graduação em Agronomia) Instituto Federal de Minas Gerais, Bambuí, 2014.

ODESTO, J. C.; RODRIGUES, J. D.; PINHO, S. Z. Efeito do ácido giberélico sobre o comprimento e diâmetro do caule de plântulas de limão Cravo (*Citrus limonia* Osbeck). **Scientia Agrícola**, Piracicaba, v. 53, n. 2/3, p. 234-240, 1996.

OLIVEIRA, A. B. et al. *Capsicum*: pimentas e pimentões no Brasil. Brasília: EMBRAPA, 2000. 113p.

OLIVEIRA, R. H. Caracterização de sintomas visuais de deficiência de micronutrientes em tomatoiro do grupo salada. **Semina**, Londrina, v. 30, v. 11, p. 1093-1100, 2006.

PEREIRA, F. E. C. et al. Physiological quality of pepper seeds as a function of age and postharvest resting time of the fruits. **Revista Ciência Agronômica**, Campinas, v. 45, n. 3, p. 737-744, 2014.

PETRI, J. L. et al. **Growth regulators for temperate climate fruit trees**. Epagri, 2016, 141 p.

POLACINSKI, J. F. **Vegetative growth of tomato plants as a function of biostimulant application**. 2020. 38 f. Monograph. (Graduation in Agronomy). Universidade do estado do Mato Grosso, Alta Floresta, 2020.

POZZOBON, M. T. et al. Meiosis and pollen viability in advanced pepper strains. **Horticultura Brasileira**, Ilha Soleira, v. 8, p. 212-216, 2011.

PRECIADO, G. M. et al; **The history of pepper.** 2010. Available at: <https://receitasmarinpreciado.wordpress.com/2010/11/12/a-historia-da-pimenta/>. Accessed on: 09 Nov. 2020.

RADKE, A. K. et al. Amino acids via seed treatment: reflections on watermelon seed vigor. **Tecnologia & Ciência Agropecuária**, Brasília, v. 11, p. 113-117, 2017.

RIBEIRO, C. S. C; HENZ, G. P. (Eds.). Sistemas de Produção, 4 - 1ª Ed. 2007. 23 p.

RODRIGUES, C. et al.; Study of the anti-inflammatory action of finger pepper (*Capsicum baccatum* L.). **Saúde e Pesquisa**, Sete Lagoas, v. 5, n. 2, p. 89-96, 2012.

RODRIGUES, D. S. et al. Amount absorbed and concentrations of micronutrients in tomato plants under protected cultivation. **Scientia Agricola**, Botucatu, v. 59, n. 1, p. 137-144, 2015.

RUFINO, J. L. S.; PENTEADO, D. C. S. Importância econômica, perspectivas e potencialidades do mercado para pimenta. **Informe Agropecuário**, São Paulo, v. 27, n. 235, p.715. 2006.

SILVA, J. V. Aproveitamento de materiais alternativos na produção de mudas de tomateiro sob adubação foliar. **Revista Ciência Agronômica**, Campinas, v. 45, n. 3, p. 528-536, 2014.

TAIZ, L.; ZEIGER, E. **Physiologia vegetal**. 3. ed. Porto Alegre: Artmed, 2004. 719 p.

TAIZ, L.; ZEIGER, E. **Plant Physiology**. 5.ed. Porto Alegre: Artmed, 2013. 954 p.

VIDIGAL, D. D. S.; DIAS, D. C. F. S.; PINHO, E. V. D. R. V.; DIAS, L. A. D. S. Physiological and enzymatic changes during maturation of pepper (*Capsicum annuum* L.) seeds. **Revista Brasileira de Sementes**, Brasília, v. 31, n. 2, p. 129-136, 2009.

VIEIRA, E. L.; CASTRO, P. R. C. **Ação de bioestimulante na cultura da soja (*Glycine max* L. Merrill)**. Cosmópolis: Stoller do Brasil. 2004. 47p.

WAGNER, A. et al. Ácido giberélico no crescimento inicial de mudas de pessegueiro. **Horticultura Brasileira**, Ilha Solteira, v.23, p. 235-243, 2007.

ZIMMER A. R et al.; Antioxidant and anti-inflammatory properties of Capsicum baccatum: from traditional use to scientific approach. **Journal Ethnopharmacol**, Amsterda, v. 139, p. 228-233, 2012.

Chapter 4

FOLIAR FERTILIZER AND BIOSTIMULANT ON BELL PEPPER PLANTS

Daiane Corrêa, Lenon Kurek and Suelen Cristina Uber

Introduction

The bell bell pepper (*Capsicum annuum* L.) is a vegetable belonging to the Solanaceae family, with origin in the tropical regions of America, including Mexico, Central America and South America (MAROUELLI; SILVA, 2014). This vegetable has great economic importance and nutritional value, because it is a source of natural antioxidants such as vitamin C, carotenoids and vitamin E, besides containing B-complex vitamins and vitamin A (REIFSCHNEIDER, 2000).

In Brazil, chili is a vegetable of great economic importance and is planted and consumed throughout the national territory. The estimated planting area is 19 thousand hectares, with production above 420 thousand tons (FAO, 2017). The bell pepper in the ranking of revenue of vegetables figured as the seventh placed, with a value of R$ 574 million (REVISTA RURAL, 2019).

The crop develops in environments with temperatures up to 30ºC, which favor the best plant development, flowering and fruit set. Very low or very high temperatures interfere with the germination and development of the plants from emergence to flowering (MAROUELLI; SILVA, 2014).

The production of seedlings in horticulture has been considered a primordial activity for most species, aiming to obtain greater uniformity of plants. However, vegetable seedlings have been produced in several ways and the current trend is to improve the means of production and improve their quality, especially with the introduction of new management techniques (ARAÚJO et al., 2000).

Foliar fertilizers and biostimulants, which are formulations that can contain mineral and organic compounds, as well as the addition of plant hormones, promote and facilitate nutrient uptake, contributing to increased growth and stress resistance in plants (VAN OOSTEN et al., 2017; POLACINSKI, 2020).

The application of foliar fertilizer, plant stimulants and/or growth regulators, aiming to improve the standards of production and productivity, has provided promising and significant results, especially in regions where the crops have a high level of technology and management (VIEIRA; CASTRO, 2004).

Thus, foliar fertilizers and biostimulants emerge as an alternative for the production of seedlings, especially in regions with higher temperatures, aiming to standardize the plants in these environments, in order to provide better growth. In this context, the objective of this work was to evaluate the effect of foliar fertilizer and biostimulant on the development of bell pepper plants.

Material and Methods

The experiment was conducted at Universidade do Estado do Mato Grosso (UNEMAT), Campus Universitário de Alta Floresta-MT *Campus2* (9°53'49.1" S 56°05'37.1" O) in the period from December 2020 to January 2021. The plants were conducted during germination, emergence, and initial development in a greenhouse with an average temperature of 27°C and daily irrigation. The climate of the region is type Am, according to Koppen's classification, being rainy tropical, reaching a high rainfall index in summer reaching values higher than 2,500 mm, and a dry winter, with predominantly high temperatures. The average annual temperature is around 25 °C (OLIVEIRA, 2006). The experimental design adopted was entirely randomized (DIC), with 5 treatments with increasing doses of foliar fertilizer and biostimulant. For seedling production, there were four repetitions with 25 experimental units, totaling 100 plants per treatment. The treatments were composed of different doses of foliar fertilizer and biostimulant (Biozyme), being 2ml/L, 4 ml/L, 6 ml/L, 8 ml/L and the control (0 ml/L), composed of water only. Each dose of foliar fertilizer and biostimulant was adjusted with the help of a syringe and inserted into a beaker containing 1 L of water.

Table 1 - Chemical composition with the water-soluble mineral elements, constituents of the foliar fertilizer and biostimulant used in the experiment.

Composition	% w/w	w/v (g/L)
Nitrogen	1,00	18
Potassium	5,00	60
Boron	0,08	0,96
Iron	0,40	4,8
Manganese	1,00	12
Sulfur	1,00	12
Zinc	2,00	24
Organic Carbon	3,50	42

Source: Arysta Life Science (2020).

The seeds of bell bell pepper Yolo Wonder were purchased commercially, in the city of Alta Floresta-MT and sown in polyethylene trays, with 200 cells, with 2 seeds allocated in each cell, which contained commercial substrate Carolina Soil, composed of Sphagnum peat, expanded perlite, expanded vermiculite, roasted rice husk and limestone. The thinning of the seedlings is performed 5 days after emergence (DAE).

During the growth of the aerial part, two applications of foliar fertilizer and biostimulant were made without each treatment, with the different doses, the first at 8 DAE and the second at 15DAE.

The evaluations were performed at 15 and 22 days after emergence. The variables analyzed were plant height (PA) cm, root length (RC) cm, total plant length (TLC) cm, stem diameter (CD) cm, relative chlorophyll content (RCT) % and number of leaves (LE) n. in 8 plants, randomly chosen in each repetition, the plants being discarded due to evaluations with destructive samples.

The height of the plant was measured by obtaining the height from the neck of the plant (cm) to its extremity, with the help of a millimeter ruler.

Root length (cm) was determined, using a millimeter ruler, from the base of the collar to the tip of the largest root, after the roots were washed in water with constant agitation, to remove the substrate that was aggregated in the root system.

With the plant height and root length data, the CTP was estimated, through the sum of the results obtained in the AP and CR evaluations.

The diameter of the stem was evaluated using a digital pachymeter in (cm), which is inserted into the stem of the plant 2 mm above the surface of the substrate.

The number of leaves was estimated by numerical counting (n.) of all the leaves present on the plant. The relative chlorophyll content was determined by taking readings with a Minolta SPAD-502 chlorophyll meter (%) in the middle part of the leaf on four leaves per replicate.

The data were submitted to variance analysis and the means were compared using Tukey's test, at 5% probability, with the help of the statistical software SISVAR.

Results and Discussion

According to the variables analyzed (Table 2) in the first evaluation, a better result can be noted in the treatments with larger dosages, especially the 8 ml/L dose, showing better vegetative development, differing from the others in relation to plant height, with 8.3 cm, total plant length, with 15.2 cm, and also having the best relative chlorophyll content of 25.4, which showed no significant differences when compared to the treatment with 6ml/L in relation to TRC which showed 25.1%.

Table 2- Plant height (PA) cm, root length (RC) cm, total plant length (TPC) cm, stem diameter (CD) mm, number of leaves (LE) and relative chlorophyll content (RCT) in bell pepper plants (*Capsicum annuum* L.) at 15 days after emergence as a function of foliar fertilizer and biostimulant application.

Treatments	AP	CR	CTP	DC	NF	TRC
Witness	6,6 b	5,5 b	12,1 b	0,9 a	4,5 b	9,0 c
2 ml/L	6,8 b	6,5 a	13,3 b	1,1 a	4,7 b	13,5 c
4 ml/L	7,2 b	6,4 a	13,6 b	1,0 a	5,0 a	17,3 b
6 ml/L	7,3 b	6,4 a	13,7 b	1,2 a	5,3 a	25,1 a
8 ml/L	8,3 a	6,9 a	15,2 a	1,2 a	5,1 a	25,4 a
C.V.	14,8	15,9	10,5	15,1	10,9	16,7

Means followed by the same letter in the column do not differ, by Tukey's test, at 5% significance level.

The results obtained in the treatments, in most cases occurs in an increasing manner, attributing the effects of larger dosages, favoring a better and faster development of seedlings. It can be seen that the highest dose studied promotes considerable growth when compared to the lowest dose and the control, which can be explained as a function of the effect of the highest dose present in the treatments correlated with the best effect observed in the results.

The authors Vieira and Santos (2005) report that, in cotton, foliar fertilizers and biostimulants increase the speed of root growth, providing greater emergence, besides originating more vigorous seedlings. However, some studies show that foliar fertilizers and biostimulants interfere with the absorption of nutrients by plants, indicating that the responses to their applications depend on other factors, such as the plant species and the composition of the humic substances present in the products used, requiring more information about the true effect of these products on plant development (FERREIRA et al., 2007).

The treatment of 2 ml/L showed the lowest results in relation to the other dosages, differing in the number of leaves and chlorophyll content, however, it showed better development than the control, especially root length, with 6.5 cm, which was greater than the control, which averaged 5.5 cm. According to Santos et al. (2013), foliar fertilizers and biostimulants result in positive effects on most characteristics, and provided the best increase in root dry mass. The form of application via seed or foliar favored the increases in phytotechnical characteristics.

The diameter of the stem did not vary significantly between treatments, showing very close results, due to the short period of development between the emergence and the evaluation. Regarding the number of leaves, the highest averages were observed in the treatments with doses of 4, 6 and 8ml/L, remaining equal to each other, while the dose of 2ml/L did not promote a significant increase in relation to the control.

The application of foliar fertilizer and biostimulant also promoted an increase in the relative content of chlorophyll present in the leaves of the plants, the control obtained the lowest results, which did not differ from the dose of 2 ml/L. The 8 ml/L dose gave the highest relative chlorophyll content, together with 6 ml/L. The treatment with 4 ml/L differed from the others with 17.3%.

According to Pelissari (2012), the chlorophyll content reflects the leaf quality of plants and as a consequence of increasing this characteristic, higher photosynthetic rate occurs, which is directly related to plant growth.

The results obtained in the second evaluation, performed at 22 days after emergence, show the treatments with higher growth averages, due to the longer period of plant development between the evaluations performed. However, it showed similar results to the first evaluation. Among the factors that contributed to improved growth, it is evident that the plants had already undergone the second foliar application of foliar fertilizer and biostimulant.

For plant height, all the doses of foliar fertilizer and biostimulant were equal, with a height of 7.6 to 8.8 cm (Table 3), with the best development results when compared to the control. It can be seen that the treatments with foliar fertilizers and biostimulants resulted in positive effects.

Table 3- Plant height (PA) cm, root length (RC) cm, total plant length (TPC) cm, stem diameter (CD) mm, number of leaves (LE) and relative chlorophyll content (RCT) in bell pepper plants (*Capsicum annuumL.*) at 22 days after emergence as a function of foliar fertilizer and biostimulant application.

Treatments	AP	CR	CTP	DC	NF	TRC
Witness	7,0 b	6,0 b	13,0 c	1,0 a	5,2 a	10,1 d
2 ml/L	7.6 ab	6,6 b	14,0 b	1,2 a	5,6 a	13,9 c
4 ml/L	7,9 a	6,8 b	14,7 b	1,4 a	5,7 a	17,8 b
6 ml/L	8,0 a	6,8 b	14,8 b	1,4 a	6,1 a	25,3 a
8 ml/L	8,8 a	8,0 a	16,8 a	1,4 a	6,1 a	25,6 a
C.V.	12,5	14,1	10,2	15,1	8,5	13,8

Means followed by the same letter in the column do not differ, by Tukey's test, at 5% significance level.

Source: Prepared by the author.

Regarding root length, the dosage of 8 ml/L of foliar fertilizer and biostimulant obtained a large increase in root development, when compared to the other treatments, reaching an average of 8 cm, thus having a significant difference with the others.

It can be observed that the use of the biostimulant at a higher dosage provides greater growth of plant roots when compared to the control, thus allowing greater exploration of the soil and better potential for plant development after transplanting. According to Ferreira et al. (2007), passion-fruit plants showed greater average root length with increasing doses of biostimulant, while the lower doses promoted less root growth.

The results obtained for the variable stem diameter showed that there were no significant differences, both in the first and second evaluation, a factor correlated to the short period of plant development.

The average results of the treatments for the number of leaves were similar, with a greater number compared to the first evaluation, however, they did not show significant differences between them, with variations from 5.2 to 6.1 leaves. According to the macros and micronutrients, as well as their quantities present in the biostimulant, there may be an influence on the greater number of leaves, as well as on other variables (POLACINSKI, 2020).

According to Maia (2014), in analyses performed on the pinyon plant, nitrogen deficiency significantly reduced leaf area, number of leaves and length, diameter and mass of dry matter of the stem; length of the root system in addition to the chlorophyll content measured by the SPAD index according to the same author the omission of sulfur caused significant reduction in: SPAD index, leaf area, number of leaves length and mass of dry matter of the stem; length, volume and mass of dry matter of roots.

Among these factors, one can explain the difference between the treatments and the control, which is evident in the relative chlorophyll content, where the control mean values reached 10.1%, while the lowest value of the treatments with a dose of biostimulant was 13.9% in the treatment with a dose of 2 ml/L, and the highest value, 25.6%, was obtained with the application of 8 ml/L of foliar fertilizer and biostimulant.

Therefore, foliar fertilizers and biostimulants are substances that interfere in plant growth. In this context, it can be observed that increasing the volume of the dosage up to 8ml/L, obtained considerable positive results, providing larger seedlings, with a root system that provides a better grip and development after transplanting, among other variables such as chlorophyll content, which contributes directly to the vegetative quality of the plant.

Conclusion

Given the results presented, it is evident that the use of foliar fertilizer and biostimulant provided advances in the development of bell pepper plants.

The treatment with 8 ml/L of Biosyme, promoted the best result for bell pepper plants, increasing their vegetative growth.

REFERENCES

ABCSEM. 2009. Available at:<http://www.abcsem.com.br/>. Accessed on: 23 Jan. 2021.

ALBUQUERQUE, F.S. et al. Lixiviação de potássio em um cultivo de pimentão sob lâminas de irrigação e doses de potássio. **Revista Caatinga**,Fortaleza, v. 24, n. 3, p. 135-144, 2011.

ALMEIDA, G.; M. et al. Plant development through interference of auxins, cytokinins, ethylene and gibberylins. **BrazilianJournalof Applied Technology for Agricultural Science**, Guarapuava, v.9, n.3, p.111-117, 2016.

ANDRIOLO, J.L. Fisiologia da produção de hortaliças em ambiente protegido. **Horticultura Brasileira**, Brasília, v.18, suppl, p.26-32, 2000.

ARAÚJO, J. A. C., CORTEZ, G. E. P., FERNANDES, C. **Efeitos dos diferentes substratos na produção de mudas de pimentão.** Jaboticabal: FCAV/UNESP. 48 p., 2000.

ARYSTA LIVE SCIENCE DO BRASIL. **Biozyme**. Available at: https://www.upl-ltd.com/br/defensivos-agricolas/tratamento-de-sementes/biozyme. Accessed on: Jan 23, 2021.

BINSFELD, J. A. et al. Use of bioactivator, biostimulant and nutrient complex in soybean seeds. **Pesquisa Agropecuária Tropical**, Brasília, v. 1, n. 1, p. 88-94, 2014.

BOARETTO, A. E. et al. **Foliar fertilization**. Campinas: Cargill Foundation, v. 2, 652 p., 1989.

BOURSCHEIDT, C. E. Bioestimulante e seus efeitos agronômicos na cultura da soja (*Glycinemax*). **Pesquisa Agronômica Brasileira**, Ijuí, v. 28, p. 201-217, 2011.

BRANDÃO FILHO,J. U. T. et al. **Fruit vegetables**. Maringá:Eduem, 2018. 535 p.

CASTRO, P.R.C.; VIEIRA, E.L. **Aplicações de reguladores vegetais na agricultura** tropical. Guaíba: Agropecuária, 2001.

CARVALHO, N. M.; NAKAGAWA, J. **Sementes: ciência, tecnologia e produção**. 4 ed. Jaboticabal: Funep. 2000. 588 p.

CARVALHO, J. A. et al. Análise produtiva e econômica do pimentão-vermelho irrigado com diferentes lâminas, cultivado em ambiente protegido. **Revista Brasileira de Engenharia Agrícola e Ambiental**, Campinas, v.15, n.6, p.569-574, 2011.

CARVALHO, L. E. et al. **Alternative practices for management of *phytophthora* wilt (*Phytophthoracapsici*) in bell pepper culture**. Bambuí: IFMG.
2015. 14 p.

DABADIA, A. C. A. Uso de bioestimulante na assimilação do nitrato e nos caracteres agronômicos em feijoeiro. **Cultura Agronômica**, Ilha Solteira, v.24, n.4, p. 321-332, 2015.

DANTAS, A. C. V. L. et al. Efeito do ácido giberélico e de bioestimulante Stimulate® no crescimento de tamarindo. **Revista Brasileira de Fruticultura**, Jaboticabal, v. 34, n. 1, p. 08-14, 2012.

DOURADO NETO, D. et al. Application and influence of phytoregulator on corn plant growth. **Revista da Faculdade de Zootecnia, Veterinária e Agronomia**. São Carlos, v.11, p.93-102, 2004.

Empresa Brasileira de Pesquisa Agropecuária- EMBRAPA. **Cultivo de tomate para industrialização**. Brasília: EMBRAPA, v. 1, 2003. 13 p. (Technical Circular).

Empresa Brasileira de Pesquisa Agropecuária. EMBRAPA, **Hortaliças Sistemas de Produção**. Electronic version, 2007. Pimenta (*Capsicum* spp.).

Empresa Brasileira de Pesquisa Agropecuária- EMBRAPA. **Cultivo de Tomate**. Brasília: EMBRAPA, v. 3, 2010. 28 p. (Technical Circular).

FAYAD, J. A. et al. Absorption of nutrients by tomato cultivated under field and protected environment conditions. **Revista Horticultura Brasileira**, Brasília, v. 20, p. 90-94, 2002.
FERREIRA, A. G., BORGHETTI, F. **Germinação: do básico ao aplicado**. Porto Alegre: Artimed. 323 p., 2004.
FERREIRA, L. A. et al. Bioestimulante e fertilizante associados ao tratamento de sementes de milho. **Revista Brasileira de Sementes**, Brasília, v. 29, n. 2, p. 80-89, 2007.

FERREIRA, M. M. M. et al. Eficiência da adubação nitrogenada do tomatoiro em duas épocas de cultivo. **Revista Ceres**, Viçosa, v. 57, n.2, p. 263-273, 2010.

FILGUEIRA, F. A. R. **Novo Manual de Olericultura: agrotecnologia moderna na produção e comercialização de hortaliças**. Lavras: Editora UFLA, 333 p., 2008.

FINGER, F. L.; SILVA, D. J. H. Cultura do pimentão e pimentas. In: FONTES, P. C. R. (Ed.). **Olericultura: teoria e prática**. Viçosa, MG: UFV, p. 429-437, 2005.

FAO, Food and Agriculture Organization of the United Nations. FAO Statistical Programme of Work. 2020. http://faostat3.fao.org/browse/Q/*/E

GENUNCIO, G. C. et al. Production of tomato cultivars in hydroponics and fertigation under nitrogen and potassium ratios. **Horticultura Brasileira**, Brasília, v. 28, p. 446-452, 2010.

GOTO, R., and SILVA, E.S. **Production of tomato, pepper and cucumber seedlings.** Maringá: EDUEM, 2018, pp. 387-400.

HF BRAZIL. HORTIFRUTI/CEPEA: **Maincharacteristics of peppers in BR.** 2017. Available at: < https://www.hfbrasil.org.br/br/hortifruti-cepea-principais-caracteristicas-do-pimentao-no-br.aspx>. Accessed on: 23 Jan. 2021.

HORNKE, N. F. et al. Use of foliar fertilizer on the physiological performance of Moranga pumpkin seeds. **Tecnologia e Ciência Agropecuária**, São Paulo, v. 12, n. 3, p. 31-35, 2018.

HUBER, D. M. The role of mineral nutrition in defense. **Academic Press**, New York, v.5, p.381-406,1980.

KANAI, S. et al. Depressionofsinkactivity precedes the inhibitionofbiomassproduction in tomatoplantssubjectedtopotassiumdeficiency stress. **Journalof Experimental Botany**, Amsterdan, v. 58, p. 2917-2928, 2007.

LANA, A. M. Q. et al. Application of growth regulators in bean culture. **BioscienceJournal.** Uberlândia, v. 25, n. 1, p. 13-20, 2009.

LIMA, A. A. et al. Foliar nutrient concentration and productivity of tomato cultivated under different substrates and doses of humic acids. **Revista Horticultura Brasileira**, Brasília, v. 29, p. 63-69, 2011.

LORENTZ, L. H.; **Variability of bell pepper fruit production in a plastic greenhouse related to experimental techniques**. 2004. 56 f.Dissertation (Master in Agronomy). Federal University of Santa Maria, Santa Maria, 2004.

MATOS, E. S.; SHIRAHIGE, F. H.; MELO, P. C. T. Desempenho de híbridos de tomato de crescimento indeterminado em função de sistemas de condução de planta. **Revista Horticultura Brasileira**, Brasília, v. 30, n. 2, p. 240-245, 2012.

MACHADO, A. Q. et al. Produção de tomate italiano (saladete) sob diferentes densidades de plantio e sistemas de prune visando ao consumo *in natura*. **Revista Horticultura Brasileira,** Brasília, v. 25, p. 149-153, 2007.

MAIA, J. T. L. S. et al. Nutrient omission in jatropha plants grown in nutrient solution. **Revista Ceres**, Lavras, v. 61, n. 5, p. 723-731, 2014.

MALAVOLTA, E. **Nutrição mineral de plantas**. São Paulo: Ed. Agr. Ceres, 631 p., 2006.

MAROUELLI, W. A. **Tensiometers for irrigation control in vegetables.** Brasília: Embrapa Hortaliças. Technical circular, n. 57, 15 p., 2008.

MARSCHNER, H. **Mineral nutritionofhigherplants**. London: Academic Press, 1995.

MARTINS, G. Cultivo em ambiente protegido. Viçosa: UFV, 148 p., 2000.

MELLO, S. C.; C. VITTI, G. C. Desenvolvimento do tomatoiro e modificações nas propriedades químicas do solo em função da aplicação de resíduos orgânicos, sob cultivo protegido. **Revista de Horticultura Brasileira**. Brasília, v. 20, n. 2, p. 275-284, 2009.

MELO, P.C.T. et al. Performance of tomato cultivars in organic system under protected cultivation. **Revista Horticultura Brasileira**, Brasília, v. 27, p. 553-559, 2009.

MOGOR, A. F. et al. Chlorophyll contents in tomato cultivars submitted to foliar applications of magnesium. **Pesquisa Agropecuária Tropical**, Goiânia, v. 43, n. 4, p. 363-369, 2013.

MOREIRA, G.R.M. et al. **Espécies e variedades de pimenta**. Belo Horizonte: EPAMIG, 2006. v.27, p.16-29. (Informe Agropecuário).

MUELLER, S. Modes of phosphorus application for two tomato cultivars. **Revista Horticultura Brasileira**. Brasília, v. 33, p. 356-361, 2015.

NAIKA, S. **A cultura do tomate**: **produção, processamento e comercialização**. Florianópolis: UFSC, 2005. 105 p.

NASCIMENTO, W. M.; DIAS, D. C. F. S.; FREITAS, R. A. Produção de sementes de pimentas. **Informe agropecuário,** São Paulo, v. 27, n. 235, p. 30-39, 2006.

NASCIMENTO, W. M. **Produção de Sementes de Hortaliças.** v. 1, 14 ed. 2014.

OLIVEIRA, R. H. Caracterização de sintomas visuais de deficiência de micronutrientes em tomatoiro do grupo salada. **Semina**, Londrina, v. 30, v. 11, p. 1093-1100, 2006.

PELISSARI, G. et al. Hormones growth regulators and their effects on morphological parameters of forage grasses. In: SEPE - Symposium on Teaching, Research and Extension. **Annals...** Unifra, 2012, Santa Maria - RS, 2012.

PETRI, J. L et al. **Growth regulators for temperate climate fruit trees**. Epagri, 2016.141 p.

POLACINSKI, J. F. **Vegetative growth of tomato plants as a function of biostimulant application**. 2020. 38 f. Monograph. (Graduation in Agronomy). Universidade do estado do Mato Grosso, Alta Floresta, 2020.

REIFSCHNEIDER, F. J. B. (Org.). *Capsicum, chilis and peppers in Brazil.* Brasília: Embrapa CNPH, 2000. 113 p.

REIS, L. S. et al. Evapotranspiration and crop coefficient of kaki tomato grown in a protected environment. **Revista Brasileira de Engenharia Agrícola e Ambiental**, Campina Grande, v. 13, p. 289-296, 2009.

REVISTA RURAL, The Industry Magazine. **Sustainable cultivation of peppers ensures producer profit.** 2019. Available at:<https://www.revistarural.com.br/2019/07/22/cultivo-sustentavel-do-pimentao-garante-lucro-aoprodutor/#:~:text=A%20hortali%C3%A7a%20%C3%A9%20cultivated%20naseja%2C0% 20113.844%2C7%20tonnes.>. Accessed on: 23 Jan. 2021.

RODRIGUES, D. S. et al. Amount absorbed and concentrations of micronutrients in tomato plants under protected cultivation. **ScientiaAgricola**. Botucatu, v. 59, n.1, p.137-144, 2002.

SMIDERLE, O.J. et al. Production of lettuce, cucumber and bell pepper seedlings in substrates combining sand, soil and Plantmax®. **Horticultura Brasileira**, Brasília, v.19, n.3, p.253-257, 2001.

SANTOS, V. M. et al. Uso de bioestimulantes no crescimento de plantas de *Zeamays* L. **Revista Brasileira de Milho e Sorgo**. São Paulo,v. 12, n. 3, p. 307-318, 2013.

TAIZ, L.; ZEIGER, E. **Plant Physiology**. 5 ed. 954 p. Porto Alegre: Artmed, 2013.

TREICHEL, M.; CARVALHO, C.; BELING, R. R. **Anuário brasileiro de sementes**. Santa Cruz do Sul: Gazeta Santa Cruz, 2016. 72 p.

VAN OOSTEN, M. J. et al. The role ofbiostimulantsandbioeffectors as alleviatorsofabiotic stress in cropplants. **Chemical and Biological Technologies in Agriculture**, Londres, v. 4, n. 5, p. 238-246, 2017.

VIEIRA, E. L.; CASTRO, P. R. C. **Ação de bioestimulante na cultura da soja (*Glycine max* (L.) Merrill).** Cosmópolis: Stoller do Brasil, 2004.

VIDIGAL, D. S. et al. Physiological and enzymatic changes during maturation of pepper (*Capsicum annuum* L.) seeds. **Revista Brasileira de Sementes**, Brasília, v. 31, n. 2, p. 129-136, 2009.

Printed by Books on Demand GmbH, Norderstedt / Germany